EXCEL

PIVOT TABLES

*For Complete Beginners, Step-By-Step
Illustrated Guide to Mastering Pivot
Tables*

**Copyright © 2018 by
William B. Skates**

Table of Contents

Introduction

Hello, there future Excel Programmers! Thanks for viewing this book. This book has been designed to be your go-to book for excel programming. Excel is a spreadsheet, which means it is a computer program [software] that enables you to collect and organize data. Excel is an extraordinary example of human ingenuity. It has built into it just about any mathematical, logical or statistical function you can think of and some others as well. Using it, you can automatically produce graphs to display your data and you can sort it in any way you like.

This book takes you a step further into the wonderful world of Excel.
It builds on your knowledge of Excel and introduces you to an amazing part of this extraordinary spreadsheet. This is the **pivot table**. You'll be guided as to what this is and the features that are built into it, which make it so valuable. These features include such things

as filters, slicers, timelines, charts, and dashboards. Such terms may seem a bit strange to you now but once you have seen them in action you will be blown away by their power and ease of use. Are you ready to start? Onto chapter 1!

Chapter 1: What are Pivot Tables?

A pivot table summarizes your data and puts it in tabular form, which allows you to report easily on trends that your information shields. Pivot tables are very valuable when you have massive rows or columns of data that you need to sum and draw comparisons from.

The pivot table is far more than just a table. It is a toolbox that allows you to create reports from data buried in tables, often with thousands of entries. By using pivot tables, you sort, filter, calculate then summarize your spreadsheet data. In this useful form, you can extract pertinent information for a report.

As an example, your spreadsheet may have 115 field columns, but you may only require five of these fields for a report. Using the pivot table and its tools you can extract that data in a few short minutes.

The term pivot in pivot table arises because in a sense you rotate or pivot the table so you see it in a different way. When a pivot table is created, there is no addition to, subtraction from, or other change to the data. All that happens is a reorganization of the data to reveal useful information lying within it.

Here is a simple example. ACE Motors LTD, a car dealership, has 3 branches A, B, and C, which employ 4 salespeople Carl, Mike, Bob, and Mary. Records are kept in a spreadsheet of the sales. A small excerpt from this spreadsheet is shown below.

It turns out that there are over 100 records for October 2017 alone. The manager is keen to find out how the different branches and salespeople are performing.

Dealer	Make	Price	Date	Salesperson
B	Ford	$14,900.00	26 October 2017	Carl
C	Honda	$9,500.00	4 October 2017	Mike
C	Honda	$6,400.00	20 October 2017	Mike
C	Mazda	$5,900.00	2 October 2017	Mike
B	Ford	$15,400.00	25 October 2017	Carl
B	Suzuki	$14,100.00	24 October 2017	Bob
B	Ford	$9,600.00	3 October 2017	Carl
B	Suzuki	$14,900.00	12 October 2017	Carl
B	Ford	$9,300.00	6 October 2017	Carl
B	Ford	$11,900.00	19 October 2017	Carl
C	Honda	$7,300.00	10 October 2017	Mike
C	VW	$9,000.00	13 October 2017	Mike
B	Suzuki	$7,300.00	24 October 2017	Bob
C	Honda	$6,600.00	24 October 2017	Mike

It would be possible, without pivot tables, to find this information but it would take many minutes, if not hours.

However, if a pivot table is used, the following report only takes a few minutes to create.

Branch People	Cars Sold	Value
A	**35**	**$326600**
Mary	35	$326600
B	**27**	**$303500**
Bob	14	$157600
Carl	13	$145900
C	**38**	**$395200**
Mike	38	$395200
Grand Total	**100**	**$1025300**

This is not a very elegant report but it only took three minutes to prepare and the manager can easily see that during October 2017, 100 cars were sold, Mary sold $326,600 worth, Bob $157,000, Carl $145,900 and Mike $395,200.

At branch A, Mary was the sole salesperson and sold 35.

You may object and say that you could get this information using sorting and filtering techniques and you could do it much more quickly than the hours I said you'd need. Have a look at the next data set.

This example is provided by the dataset *P1-SuperStoreUS-2015.xlsx* , which you can download from https://www.superdatascience.com/pages/tableau. This is the web site of the SuperDataScience organization, an organization that trains data scientists.

Here is a copy of a small part of a few rows from this set, which has more than 20 columns:

Region	State or Province	City	Postal Code	Order Date	Ship Date	Profit	Quantity ordered new	Sales
West	Washington	Anacortes	98221	7/01/15	8/01/15	4.56	4	13.01
West	California	San Gabriel	91776	13/06/15	15/06/15	4390.3665	12	6362.85
East	New Jersey	Roselle	7203	15/02/15	17/02/15	-53.8096	22	211.15
Central	Minnesota	Prior Lake	55372	12/05/15	14/05/15	803.4705	16	1164.45

There are nearly 2000 rows. With a spreadsheet that is so large, doing any sort of meaningful analysis is almost impossible, when done manually using the functions of Excel. Despite this, it only took me a few minutes to answer the question, "How much revenue was generated in the different regions. The answer follows in the following pivot table.

Discount(%)	Central	East	South	West	Grand Total
0	$60433.34	$16928.69	$47186.61	$45990.41	$170539.05
1	$24612.38	$76141.07	$20110.17	$45486.48	$166350.1
		$118409.1			
2	$30849.79	2	$40056.34	$61499.64	$250814.89
3	$45068.42	$64970.57	$36480.29	$43970.88	$190490.16
4	$60276.58	$37182.55	$25910.55	$53341.12	$176710.8
5	$69543.11	$55155.49	$28068.01	$65257.6	$218024.21
6	$34592.36	$37814.04	$27359.02	$36276.75	$136042.17
7	$40032.81	$55149.42	$27757.92	$84813.63	$207753.78
8	$30493.19	$41115.29	$33944.17	$27045.5	$132598.15
9	$32821.02	$52657.95	$42153.1	$42242.31	$169874.38
10	19379.87	$36647.3	$28078.94	$20852.25	$104958.36
17	$27.96				$27.96
21	$153.87				$153.87
		$592171.4	$357105.1	$526776.5	$1924337.
Grand Total	$448284.7	9	2	7	88

This table has a goldmine of information on it. It is really easy to see the revenue earned in the different regions. It is fascinating to see the effect of different discounts on revenue. In

chapter 3, we will show how to build pivot
tables like this.

Chapter 2: Fundamentals of Pivot Tables: What you can do with them?

As stated and demonstrated previously, pivot
tables are tools, which allow the user to rapidly
extract meaningful information from
spreadsheets. They allow tasks that would take
hours or be nearly impossible to be done in less
than a minute.

What sort of information, are we talking about?

Consider the previously mentioned
spreadsheet,

Region	State or Province	City	Postal Code	Order Date	Ship Date	Profit	Quantity ordered new	Sales
West	Washington	Anacortes	98221	7/01/15	8/01/15	4.56	4	13.01
West	California	San Gabriel	91776	13/06/15	15/06/15	4390.3665	12	6362.85

with nearly 2000 rows of more than 20 columns. Done manually using the functions of Excel, any sort of meaningful analysis is difficult, if not impossible. However, it only took me a few minutes to answer the question, "How much revenue was generated in the different regions?"

There are so many questions that can be answered successfully using pivot tables. Instead of asking the question, "How much revenue was generated in the different regions?", an investor might ask, "How much profit was generated in the different regions?"

The answer is shown in the following pivot table.

Discount(%)	Central	East	South	West	Grand Total
0	$17,071.27	-$5,656.86	-$1,097.97	$19,155.93	$29,472.38
1	$2,521.88	$14,020.64	-$1,539.14	$8,011.85	$23,015.24
2	$7,774.81	$22,856.04	-$13.40	$12,099.34	$42,716.79
3	$18,169.98	$20,469.46	$221.41	$21,025.32	$59,886.17
4	-$4,734.39	$5,776.28	-$755.83	-$4,951.66	-$4,665.60
5	$15,363.28	$13,803.26	$1,036.01	$5,906.42	$36,108.97
6	$7,856.96	$12,138.25	$928.92	$10,101.63	$31,025.76
7	$9,190.09	-$2,344.80	-$2,260.77	$8,374.64	$12,959.16
8	$5,277.13	$14,939.53	-$14,555.46	$1,665.13	$7,326.34
9	$4,930.19	-$11,937.74	-$30.51	-$10,986.77	-$18,024.83
10	-$6,028.86	$1,227.33	$3,642.68	$5,442.96	$4,284.11
17	-$9.13				-$9.13
21	-$17.75				-$17.75
Grand Total	$77,365.47	$85,291.40	-$14,424.05	$75,844.79	$224,077.61

What other gems could we extract from this spreadsheet? The following questions are only a fraction of those that could be asked and answered using pivot tables.

1) What is the number of sales for each region?

Discount(%)	Central	East	South	West	Grand Total
0	46	37	36	47	166
1	53	44	41	50	188
2	55	48	48	38	189
3	52	43	46	51	192
4	65	38	37	40	180
5	58	59	33	43	193
6	50	42	38	48	178
7	53	46	38	40	177
8	42	30	42	36	150
9	48	45	46	43	182
10	42	42	37	34	155
17	1				1
21	1				1
Grand Total	**566**	**474**	**442**	**470**	**1952**

2) What were the quantities ordered new for each region at different discounts?

Discount(%)	Central	East	South	West	Grand Total
0	651	424	391	733	2199
1	752	700	426	585	2463
2	670	788	637	468	2563
3	530	704	530	532	2296
4	942	462	499	589	2492
5	593	884	393	729	2599
6	448	560	372	442	1822
7	544	755	399	758	2456
8	604	444	443	630	2121
9	469	636	481	680	2266
10	361	674	474	471	1980
17	1				1
21	10				10
Grand Total	6575	7031	5045	6617	25268

3) Which day of the week generates the most sales?

	Monday	Tuesday	Wednesday	Thursday	Friday	Saturday	Sunday	Grand Total
T	285	249	290	297	281	314	236	1952

4) Which 10 customers generated the highest profits?

Customer	Profit
Richardson McClure LTD	$9,300.34
Annie Odoms	$9,243.26
Christine Meadows	$8,839.23
Nancy Kelly	$8,533.43
Kellyanne Bryers	$7,495.06
Stein Bakery	$7,257.76
Davies Electrical	$7,024.21
Scott McKenna Software	$6,610.20
Andrea Shaw	$6,106.73
T H Wong and Associates	$5,998.03

5) Which 10 customers generated the worst losses?

Customer	Loss
Ellison Design	-$3,666.57
Scott Bunnen Consolidated	-$3,839.84
Kendally Mining	-$3,971.06
Glenda Hunterman	-$4,019.85
Luis Kerrison	-$4,118.16
Gregory Cranston	-$4,793.00
Sidney Austen Consulting	-$6,991.09
Vicki Worral Hairdresser	-$13,535.45
James Ballinger Plumber	-$13,641.55
Derek Juarez and son	-$13,706.46
Deana Solomona Enterprises	-$16,585.11

6) What methods of shipping were used, how often and in what states?

State	Delivery Truck	Express Air	Regular Air	Grand Total
Alabama	4	3	15	22
Arizona	6	3	22	31
Arkansas	2	2	18	22
California	29	28	157	214

7) What were the 10 types of merchandise sold most frequently?

Item	Times sold
Bevis 36 x 72 Conference Tables	9
80 Minute Slim Jewel Case CD-R, 10/Pack - Staples	8
Avery 493	8
Office Star - Mid Back Dual function Ergonomic High Back Chair with 2-Way Adjustable Arms	8
Lock-Up Easel 'Spel-Binder'	7
Newell 323	7
Newell 343	7
3M Organizer Strips	6
Accessory41	6
Avery 494	6

8) Which states produce the most Profit?

State	Profit
California	$37,421.96
Texas	$28,078.85
New York	$27,611.94
Ohio	$23,410.84
Oregon	$17,931.04

As you can see, it is possible to answer a very large number of questions.

The next chapter shows how to begin using pivot tables. It also shows a few 'quirks' that the beginning user of this remarkable tool should be aware of.

Chapter 3: Creating your first Pivot Table

The first step in creating a pivot chart is to get a spreadsheet with data you wish to analyze.

I am going to do all the demonstrations in this chapter with the following spreadsheet. You can download this file from

www.dropbox.com/s/rwtfao8dnl2tyc9/
SampleData_2.csv?dl=0

Item	Carat	Colour	Clarity	Shop	Value	Date of Sale	Day
Ring	0.33	6	1	B	$768.6	17/09/13	Monday
Necklace	0.31	5	1	C	$788.2	26/11/14	Tuesday
Earring	0.31	6	1	B	$788.2	18/05/14	Saturday
Broach	0.32	4	1	A	$841.4	20/09/14	Friday

There are actually more than 200 rows in this spreadsheet. First, a few comments about it. Carat is a term used in jewelry to denote the weight of gemstones. Most of the other columns are self-evident, but some may be confused by *Day*. It refers to the day of the week of the date. So, according to the spreadsheet, 26/11/14 was a Tuesday.

Using this spreadsheet and pivot table, I will show how to answer the following 6 questions:
1. Which day produced the most sales?
2. Which item generated the most valuable sales?
3. Which shop sells the greatest value?
4. Which shop sold the most items?
5. Which shop sold the most rings?
6. What was the most valuable item sold?

This analysis will use the more recent versions of Excel. Although some earlier versions of Excel had pivot tables, they were not nearly as user-friendly as the more recent versions of Excel.

Now to start; having loaded the spreadsheet into Excel go to **Insert** if using a PC or later version of Mac Excel or **Data**, if using Excel for Mac 2011.

Look for the following symbol.

PivotTable

Click it.

Here's what happened on my computer, which is a Mac.

Some very interesting little windows came up.

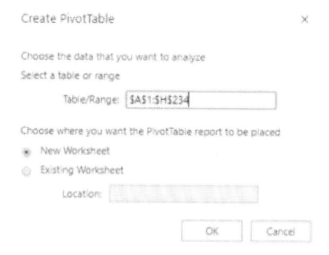

I selected a new worksheet and you should always do this, rather than have the pivot table on an old sheet.

Originally my data was on Sheet 2, which was the only sheet in the workbook, and the sheet with the pivot table was Sheet 3.

The first stage in the creation of a pivot table based on the jewelry spreadsheet had taken place.

Obviously, if you're using a different spreadsheet you will get different results. Check this yourself!

On the new sheet, you will find this, although more spread out.

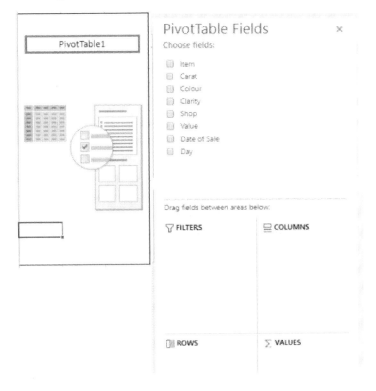

Looking at the pivot table, there is now no information showing. This is all very interesting but we still haven't answered the question, " Which day produced the most sales?".

I am going to use the pivot table fields tool [right window] to create a pivot table that answers this.

At any stage, it is always a good idea to think about what sort of table would answer the question. I think you would agree that the table below does so perfectly.

Days	Sales
Saturday	42
Friday	40
Thursday	38
Monday	33
Wednesday	32
Sunday	25
Tuesday	23
Grand Total	**233**

My job, now, is to show you how to do this.

Before, we begin, it is necessary to see what happens if you click outside the pivot table.

If I did so then this was what I got on the sheet [Sheet 3] with the pivot table.

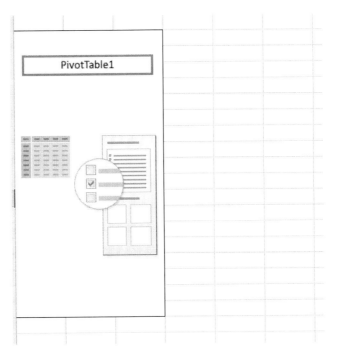

No sign of the pivot table fields.

Actually to get them back is very easy.

Click on the actual pivot table and the pivot table fields come back.

Now bear in mind the table we wanted.

Days	Sales
Saturday	42
Friday	40
Thursday	38
Monday	33
Wednesday	32
Sunday	25
Tuesday	23
Grand Total	**233**

This table tells you what to do next.

Go to the pivot tables, grab *Days* and move them to the *Rows* square.

When you do so, you get this.

Row Labels ▾	
Sunday	
Monday	
Tuesday	
Wednesday	
Thursday	
Friday	
Saturday	
Grand Total	

Notice how the days of the week have appeared in the left column, just like you want.

Now for the right-hand side of the Pivot Table.

Go to the fields list and grab *Value* and move them to the *Values* square.

When you do this, you get the following.

Row Labels ▾	Sum of Value
Sunday	72747.5
Monday	89368.3
Tuesday	67266.5
Wednesday	108340.4
Thursday	114859.5
Friday	124008.5
Saturday	109081
Grand Total	**685671.7**

We wanted the number of sales not the value of sales. Before making the correction needed, take a look at the fields list now.

PivotTable Fields

Choose fields:

- ☐ Item
- ☐ Carat
- ☐ Colour
- ☐ Clarity
- ☐ Shop
- ☑ **Value**
- ☐ Date of Sale
- ☑ **Day**

We are now using Value and Day so they now have ticks.

Let's now get the number of sales, which is a count.

Go back to the Fields List and click on Sum of Values.

When you do this window appears.

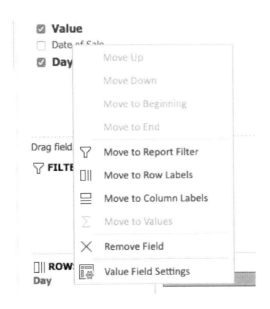

Now click on Value Field Settings.

Another window appears.

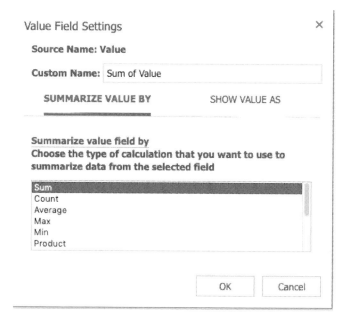

Change Sum to Count. While you're doing this have a look at the other calculations you can perform.

Anyway, as you change Sum to Count the table you want appears.

Although it would be adequate, it can easily be improved.

One thing you can do now and that is to put the data in descending order of the number of sales.

Click on the arrow at the end of the phrase **Row Labels**. When you do a window appears.

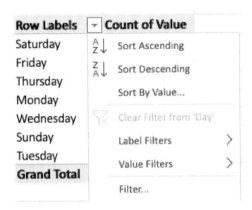

As we wish to sort by number of Sales, select Sort By Value. Yet another window appears.

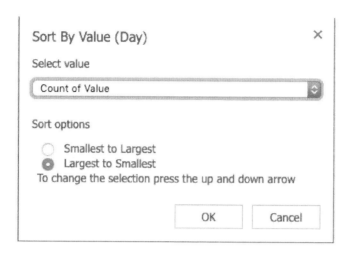

As we want to descend, we pick Largest to Smallest.

Once this is done, we get

Row Labels	Count of Value
Saturday	42
Friday	40
Thursday	38
Monday	33
Wednesday	32
Sunday	25
Tuesday	23

Grand Total 233

If you want to make a proper report then copy the entire pivot table and paste it onto a Word document. Once this is done, amending it is quite simple, using the techniques for editing tables in Word.

Days	Sales
Saturday	42
Friday	40
Thursday	38
Monday	33
Wednesday	32
Sunday	25
Tuesday	23
Grand Total	**233**

The first question was, " Which day produced the most sales?" The answer is obviously on

Saturday. Weekends would give people more people to shop for jewels.

We have now answered the first question, now onto the next, which is, "Which item that generated the most valuable sales?"

Before starting, decide which field will go on the left and provide rows. In this case, we would probably have *Item*.

Let's see if this is right by using the Fields List and dragging *Item* down to *Rows*.

The Pivot Table is now starting to look like a useful display.

Row Labels	Count of Value
Broach	59
Earring	1
Earrings	30
Necklace	59
Ring	84
Grand Total	233

As we want Value and not Count, we click on *Count of Value* again and change it to *Sum of Value*.

Row Labels	Sum of Value
Broach	178852.1
Earring	788.2
Earrings	87671.5
Necklace	184041.2
Ring	234318.7
Grand Total	685671.7

There is a problem, the numbers are just that. They are not currency. This is a really easy problem to solve. Just select the whole column, which contains the *Sum of Value* and format it to currency.

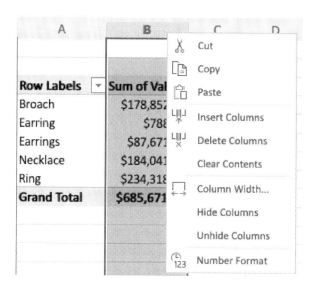

Then

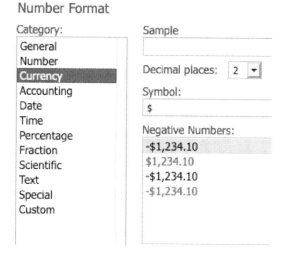

and finally

Row Labels ▼	Sum of Value
Broach	$178,852.10
Earring	$788.20
Earrings	$87,671.50
Necklace	$184,041.20
Ring	$234,318.70
Grand Total	**$685,671.70**

This is now sorted, as previously, to get

Row Labels	Sum of Value
Ring	$234,318.70
Necklace	$184,041.20
Broach	$178,852.10
Earrings	$87,671.50
Earring	$788.20
Grand Total	$685,671.70

The second question has now been answered. Rings generated the most valuable sales.

Before answering another question, we need to have a look at the sort of values available to us. So far, we have used *count* and *sum*.

Return to the Fields List and click on *Sum of Value*

The little window pops up giving a list of different mathematical functions to use.

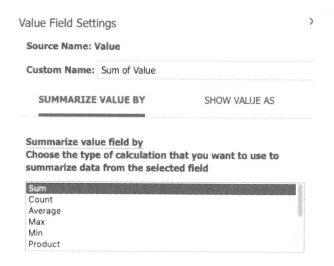

Some like *Count* could be used on any field but others like *Sum* and *Average* are only applicable to numerical fields, such as [in this case] *Value, Clarity* and *Carat.*

We will answer one more question from the six that were given at the start of this chapter. Hopefully, you will download the Excel file given earlier and try to answer the others yourself.

The final question for this chapter is, " What was the most valuable item sold?

This time, the Row Labels will again be *Item*, and we will not need any Column Labels. The Values will be Max of Value.

The table we get is

Row Labels	Max of Value
Broach	$11,205.60
Necklace	$10,907.40
Ring	$7,993.30
Earrings	$6,897.10
Earring	$788.20
Grand Total	$11,205.60

If we arrange it in descending order.

The most valuable item was a broach valued at
$11,205.60.
One of the questions asked at the chapter's
beginning was, " Which shop sold the most
rings?" This question is best addressed by what
is called *Filters*. This is the title of the next
chapter.

Chapter 4: Filters

*A filter for something is a device for removing
things that pass through it or are already
there.* A definition of a filter for a pivot table is
very difficult to find. There are many sites that
tell you how to install filters on your pivot table
but I could not find a definition for one.

Here is a possible definition, "A filter for a pivot table is a means of selecting variables of interest." You will notice that by this definition, a pivot table is itself a filter on the data of the spreadsheet. However, filters for pivot tables need to be able to get down to ranges of items or individual items of the fields [columns] of the spreadsheet.

I will show how this is done using the spreadsheet that I used in the last chapter. The first question that I address with a filter, "What is the top price of the rings sold?"

We have the pivot table

Row Labels	Max of Value
Broach	$11,205.60
Necklace	$10,907.40
Ring	$7,993.30
Earrings	$6,897.10
Earring	$788.20
Grand Total	$11,205.60

Constructed in the last Chapter.

So far, we have not used the Report Filter, which you can see in the Pivot Table Fields.

The use of the filter to answer the question, "What is the top price of the rings sold?" is really simple. All you have to do is drag *Item* up to the pane labeled Filter. Notice that it disappears from the *Rows* pane or square.

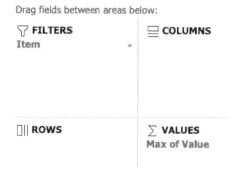

When you do this, the following pivot table is created.

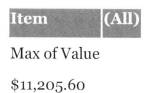

Max of Value

$11,205.60

Notice the arrow beside the word ALL. Click on it. Another window appears.

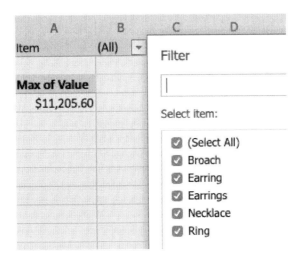

Now deselect everything except Ring.

The top price paid for a ring is $7993.30.

Suppose we clicked Necklace as well. This is what would be shown.

Filter

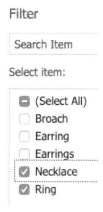

And this what we would get

Item	(Multiple Items)
Max of Value	
$10,907.40	

The conclusion would be that at least one necklace was sold for a higher price than any ring.

Now let's use multiple filters to answer the following question, "What is the most valuable necklace sold at shop C and when was it sold?"

The following pivot table provides the answer.

Item	Necklace
Shop	C

Row Labels	Max of Value
21/03/2013	$10,907.40
25/05/2014	$7,630.00
5/03/2013	$6,919.50
20/10/2013	$5,526.50
20/05/2013	$4,696.30
20/11/2013	$4,060.00
9/03/2013	$3,857.00
17/10/2013	$3,728.20
24/04/2013	$2,744.70
24/09/2013	$2,605.40
30/06/2014	$2,111.20
28/11/2014	$2,024.40
19/06/2014	$1,131.20
3/08/2014	$1,078.00
6/04/2013	$1,057.00
11/12/2014	$955.50
21/06/2014	$935.90

16/11/2014 $909.30

23/04/2013 $866.60

26/11/2014 $788.20

10/08/2014 $560.00

11/04/2013 $493.50

7/04/2014 $493.50

Grand Total $10,907.40

The answer to this question is a necklace valued at $10,907.40, which was sold on 21/03/13.

How is this pivot table created?

Here is a picture of the Fields that produced this.

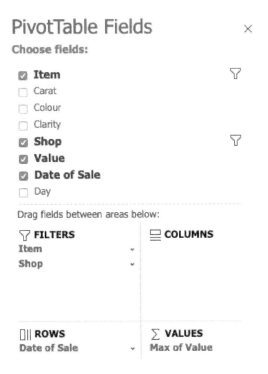

PivotTable Fields   ✕

Choose fields:

- ☑ **Item**   ▽
- ☐ Carat
- ☐ Colour
- ☐ Clarity
- ☑ **Shop**   ▽
- ☑ **Value**
- ☑ **Date of Sale**
- ☐ Day

Drag fields between areas below:

▽ FILTERS	🖵 COLUMNS
Item ⌄	
Shop ⌄	

▯‖ ROWS	∑ VALUES
Date of Sale ⌄	Max of Value

The next demonstration of a filter shows the sales of all necklaces, which were worth more than $4,000.

Item	Necklace
Shop	(All)

Row Labels	Max of Value
21/03/2013	$10,907.40
21/06/2013	$9,739.10
25/05/2014	$7,630.00
26/02/2014	$7,499.10
5/03/2013	$6,919.50
31/10/2013	$6,829.90
15/07/2014	$6,733.30
26/09/2013	$6,694.10
17/08/2014	$6,211.10
28/12/2013	$6,211.10
15/03/2014	$5,722.50
20/10/2013	$5,526.50
20/05/2013	$4,696.30
27/02/2013	$4,399.50
16/10/2014	$4,266.50
24/03/2014	$4,172.70
23/02/2013	$4,155.90
20/11/2013	$4,060.00

17/02/2014 $4,016.60

Grand Total $10,907.40

The fields required to do this are those we used to produce the answer to, "What is the most valuable necklace sold at shop C and when was it sold?"
The only difference is that the Shop filter now gives the results of all three shops and not just C.

If this is done a pivot table with more than 50 rows of values is produced. If you want to filter by value then click on the downward arrow. When this is done, the following window appears.

Item	Necklace
Shop	(All)

Row Labels	Max of Value
21/03/2013	
21/06/2013	
25/05/2014	
26/02/2014	
5/03/2013	
31/10/2013	
15/07/2014	
26/09/2013	
17/08/2014	
28/12/2013	$6,211.10
15/03/2014	$5,722.50
20/10/2013	$5,526.50
20/05/2013	$4,696.30
27/02/2013	$4,399.50
16/10/2014	$4,266.50
24/03/2014	$4,172.70
23/02/2013	$4,155.90
20/11/2013	$4,060.00
17/02/2014	$4,016.60

Menu overlay:

- A→Z Sort Ascending
- Z→A Sort Descending
- Sort By Value...
- Clear Filter from 'Date of Sale'
- Date Filters >
- Value Filters >
- Filter...

Value Filters submenu:

- Clear Filter
- Equals...
- Does Not Equal...
- Greater Than...
- Greater Than Or Equal To...
- Less Than...
- Less Than Or Equal To...
- Between...
- Not Between...
- Top 10...

As we are concerned with necklaces of value exceeding $,4000, we click by value then Greater Than.. When we do so, we get another change of window.

As we want necklaces of value greater than $4,000, we put in 4000 then click OK. When we do this, an amazing change occurs. Almost miraculously, as 4000 is entered in the textbox the pivot table shortens to the values we want.

Having thoroughly covered filters it is now time to have a good look at two more wonderful tools in Excel's bag of tricks. These are Slicers and Timelines. They are the topic of the next chapter.

Chapter 5: Slicers and Timelines

Slicers

Slicers are another method of filtering. Unlike everything else that has so far been discussed in this book, these are not available in the free online version of Excel. They are available in Office 365, which is the version of Office that you have to pay for.

The best way to show a slicer in action is to create one.

They can be created in either spreadsheets or pivot tables.

We will start with a pivot table.

Shop	(All)
Item	Ring

Row Labels	Sum of Value
10/11/2014	6261.5
24/08/2014	7484.4
20/08/2014	4843.3
4/08/2014	6701.1
1/08/2014	4763.5
27/06/2014	6151.6
1/03/2014	6442.1
27/10/2013	6927.2
14/10/2013	7993.3
2/10/2013	5376
12/07/2013	4774.7
30/06/2013	7260.4
20/05/2013	6752.2
9/05/2013	7190.4
6/04/2013	6636
3/03/2013	5472.6
Grand Total	101030.3

Let's put a slicer in it and see what is involved.

First click in any cell of the table and then on the *slicer* icon on the Insert menu

When you do this, the following little window appears.

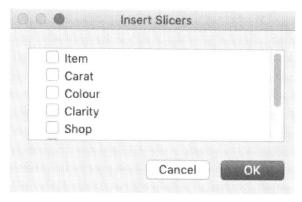

If you select *Item* then click OK then you get this. This is a Slicer.

Notice how the item we are filtering on is greyed.

Click on Earrings.

When you do this is what the slicer looks like.

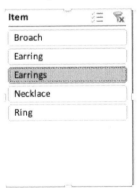

Here is how the pivot table changed.

Shop	(All)
Item	Earrings
Row Labels	Sum of Value
25/11/2014	5666.5
31/08/2014	6511.4
17/07/2014	6897.1
24/05/2014	6766.2
9/07/2013	6146.7
13/03/2013	4817.4
23/02/2013	5918.5
Grand Total	42723.8

Unlike ordinary filters, where if you pick more than one filter you just get told there are *multiple filters*, you can actually pick multiple filters and see which are in use.

If using a PC click Ring, hold down Control then click Earring. If using a Mac click Ring, hold down Command then click Earring. Here is what you get.

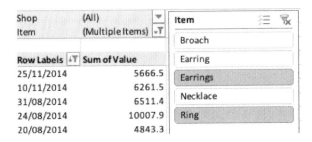

Notice how the item filter just says Multiple Items, while on the slicer both Earrings and Ring are greyed. Also, notice that the pivot table gets much greater.

Now, let's add another slicer for Shop. We do exactly as before but select Shop now.

Here is what we get

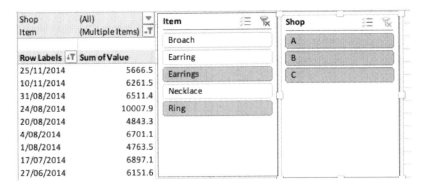

Notice that as we had all shops chosen then all shop buttons on the slicer were grey.

Let's see what happens if we only pick Ring and shop C.

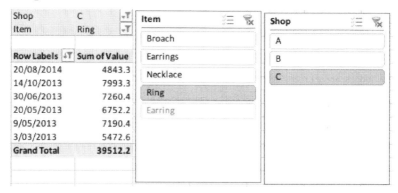

The pivot table has got much smaller and Shop C and Item Ring are now showing as filters. There are extensive references to slicers on the Internet and you are urged to read them for further help and instruction.

The next topic we look at is Timelines.

Timelines

Like *slicers*, timelines are another method of filtering, but only for time. Like slicers, these are not available in the free online version of Excel. They are available in Office 365, which is the version of Office that you have to pay for.

The best way to show a timeline in action is to create one.

We will start with a pivot table.

Shop	A
Item	Ring

Row Labels	Sum of Value
27/06/2014	6151.6
27/10/2013	6927.2
2/10/2013	5376
6/04/2013	6636
Grand Total	25090.8

Click anywhere in the pivot table then and then on the *Timeline* icon on the Insert menu

When you do this, the following little window appears.

Notice that only the field in data format is displayed.

Click Date of Sale and the following window is displayed.

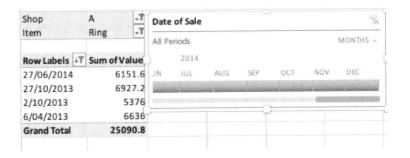

This window is a slider for the date of sale. At the moment, it shows all ring sales.

We can move it around and widen it.

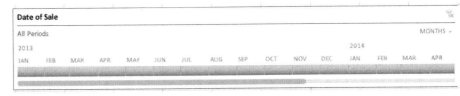

You can change the time from month to day, year or quarter.

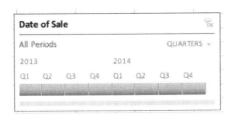

Using this let us see what the sales were in the 2nd and third quarters of 2014.

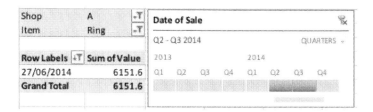

There was only one sale of $6151.60 of a ring in Shop A at that time.

Now let's see if we go to the slider and change the time to years and find sales of rings in shop A during all of 2014.

Chapter 6: Pivot Charts

It is possible to put charts or graphs of various types into pivot tables to illustrate how the data is distributed and emphasize points. As the saying goes, "one picture is worth 1000 words" and a visual display with a column graph, pie chart, bar graph or another chart can make a point that you need to get across very obvious.

Let us give an example. Suppose we have the pivot table

Shop	(All)
Count of Value	
Row Labels	Total
Monday	9
Tuesday	8
Wednesday	11
Thursday	4
Friday	9
Saturday	9

Sunday	9
Grand Total	59

and wish to display this information visually. we can do this by using a column graph. If you are using Excel OneDrive (Excel online) then there is this display on the Insert menu.

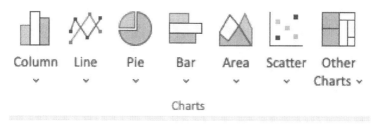

Click on Column icon, which is the one shown on the left. Once this is done, the following graph appears.

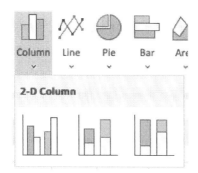

The ordinary column graph is the choice on the left.

Click on this and a very nice column graph appears between the pivot table and the Fields List.

The means by which you can access the charts will vary with the version of Excel, which you are using, however, if you have a pivot table there will be the ability to use charts with it.

Before we create a different chart, it is useful to use our filter for Shop and see what happens to the column graph.

Click on the arrow on the top of the pivot chart. When you so you get this window.

This gives you the ability to change the shop. At the moment data from all shops are being shown. Use the window, so that only data from Shop C is available.

When this is done, the following pivot table is created.

Shop	C

Count of Value

Row Labels	Total
Monday	3
Tuesday	3
Wednesday	5
Thursday	1

Friday	3
Saturday	5
Sunday	3
Grand Total	23

This table is now shown in the column graph.

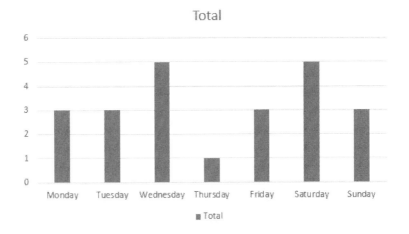

Before leaving this chapter, let us have a look at the use of another widely used chart, namely pie graphs.

Pie graphs are best used when the number of variables is 3 or 4, although they are often used with more variables. In our case, we have Item, with only 4 types. Create this pivot table.

Shop	(All)
Row Labels	Count of Value
Broach	59
Earrings	30
Necklace	59
Ring	84
Grand Total	232

Once again, go to the chart menu.

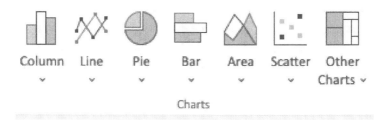

But this time pick Pie. When you do you get

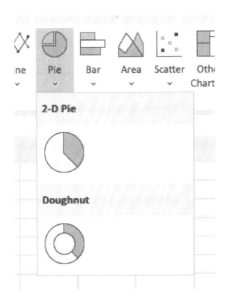

Click on the 2-D pie.

Here is what I got.

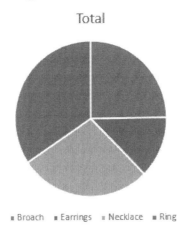

Total

■ Broach ■ Earrings ■ Necklace ■ Ring

Once again, let's see what happens when the filter is used to have Shop A.

The pivot field is now

Shop	A
Row Labels	Count of Value
Broach	9
Earrings	1
Necklace	11
Ring	7
Grand Total	28

and the pie chart is

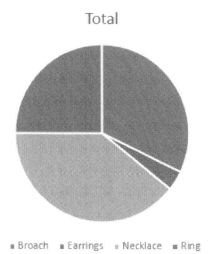

Total

Broach   Earrings   Necklace   Ring

There has been an automatic change in the chart, just like there was for Column Graphs.

We could continue showing examples, there are plenty of charts available, as you can see from the Menu.

However, the general method of using charts is always the same.

In the next chapter, we will investigate calculated fields in pivot tables.

Chapter 7: Calculated Fields

This chapter looks at yet another feature of pivot tables, which is not available on the free version of Excel but can be obtained if you have Office 365 or a recent paid version.

We will consider the pivot table

Row Labels	Sum of Value
27/10/2013	6927.2
14/10/2013	7993.3
2/10/2013	5376
12/07/2013	4774.7
30/06/2013	7260.4
20/05/2013	6752.2
9/05/2013	7190.4
6/04/2013	6636
3/03/2013	5472.6
Grand Total	58382.8

If you try and create a function within this you get an angry message from Excel.

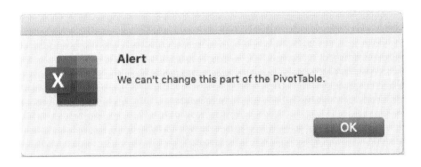

The message used to be far more aggressive in earlier versions of Excel.

In order to show calculated fields work, let's make a very simple one.

Assuming you have a version of Excel into which you can put calculated fields then right click on the pivot table select **Formulas→Calculated Field**.

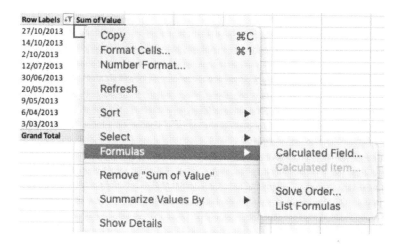

The result is this.

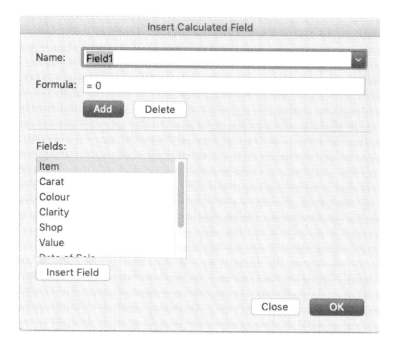

You can name the field anything you like. I am going to call it Dodo, so I type this name into the Name text field and for formula add Carat and Clarity, which are two numerical fields.

By use of the Insert Field button, I get

And then get the pivot table as

Row Labels ↓T	Sum of Value	Sum of Dodo
27/10/2013	6927.2	3
14/10/2013	7993.3	2
2/10/2013	5376	1.9
12/07/2013	4774.7	5.02
30/06/2013	7260.4	2.02
20/05/2013	6752.2	2.1
9/05/2013	7190.4	2.01
6/04/2013	6636	2
3/03/2013	5472.6	2
Grand Total	58382.8	22.05

With this field list

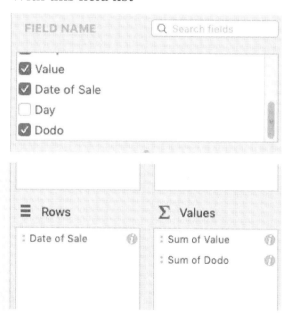

Dodo was a pretty useless calculated field and you may well wonder how you could remove it. This is quite easy.

Either bring up Calculated Field as before or find the icon shown below on the Insert ribbon then click on it.

When you do, you get

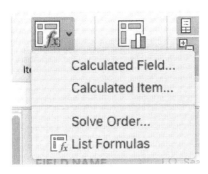

Now pick Calculated Field

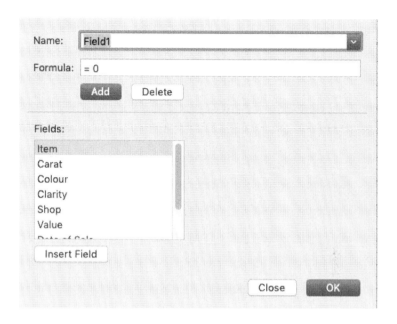

Type in Dodo and click Delete. The result is the destruction of Dodo and the restoration of the original pivot table.

Now let's create a far more useful and realistic calculate field. Often in business, sales of different size attract different commissions or bonuses to the salesperson.

In this situation let any sale over $2000 get a 5% bonus, over $1000 get a 4% bonus and anything else get 3%.

Here is how the Calculated Field window would look.

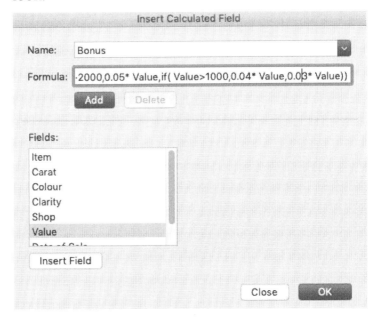

The result on the pivot table would be

Row Labels ↓T	Sum of Value	Sum of Bonus
28/12/2013	2241.4	112.07
10/12/2013	2605.4	130.27
27/10/2013	6927.2	346.36
7/10/2013	928.9	27.867
2/10/2013	5376	268.8
17/07/2013	1139.6	45.584
13/05/2013	562.1	16.863
13/04/2013	2392.6	119.63
6/04/2013	6636	331.8
Grand Total	28809.2	1440.46

Finally, we will look at what are called *Dashboards*.

Chapter 8: Dashboards

Dashboards are not really a new feature of pivot tables, but rather the gathering of the features we have talked about together.

But what is a dashboard?

The online encyclopedia Wikipedia defines dashboard as follows in reference to cars, "A dashboard is a control panel usually located directly ahead of a vehicle's driver, displaying instrumentation and controls for the vehicle's operation."

What does a dashboard in computing look like?

There are lots of dashboards in computing. Desktops are a type of dashboard. They provide a display of the computer's controls. There are windows actually called dashboards. Here is a dashboard from the Macintosh Operating System.

Creating a pivot table dashboard

Excel provides a feature called *dashboard* that allows you to gather pivot tables and charts into dashboard form.

Here is a dashboard made quite quickly for the dataset that we have been using.

There is nothing really new, except the slicers for Months and Quarters.

All the other features shown, such as the slicers for Item and Shop and the column graph for Value and Bonus can be created using techniques already shown.

Now let us discuss Months and Quarters. You may remember that one of the fields was Date of Sale. If you right click in that field you notice *Group* and *Ungroup*.

At this stage, all you have in rows is Date of Sale.

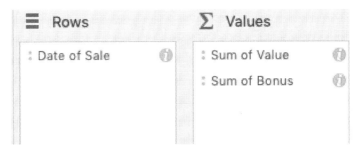

Click Group and you get

Be sure to select those time periods you want
before clicking OK. I have selected Days,
Months and Quarters. Amazingly, when you do
this Months and Quarters now appear in the
Fields List.

Slicers can now be put onto your dashboard for
Months and for Quarters. Having created this
simple dashboard, you should play around with
it and get a feel for its capabilities. You may
well be enthused to create dashboards of your
own on this or other data sets.

Chapter 9: Older Forms of Excel

Earlier, I said that the early forms of Excel prior to around 2010, did not have pivot tables. If you have such an old version remember that there is a free online version of Excel, which does have a very good form of pivot tables, although it is minus slicers, timelines, and calculated fields.

If you have Mac Excel 2011 then the form of Excel you have is about as powerful as the online version.

Here is a quick look at the creation of a basic pivot table using it, remember the pivot table icon is found in the Data ribbon [menu].

PivotTable

In Office 365, we refer to the Fields List. In the old Mac version that was called the builder and appeared as a dark grey window.

This is the table we want.

Days	Sales
Saturday	42
Friday	40
Thursday	38
Monday	33
Wednesday	32
Sunday	25
Tuesday	23
Grand Total	**233**

Go to the builder and grab *Days* and move them to the *Row Labels* black square.

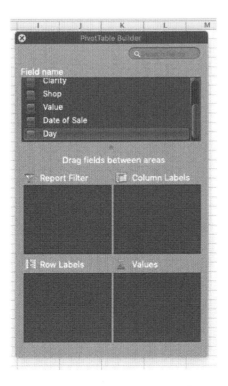

When you do so, you get this.

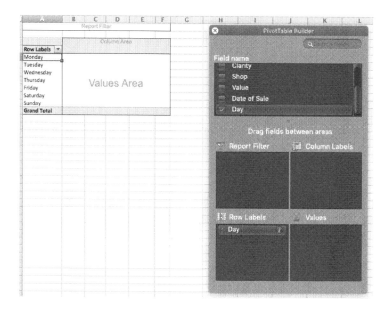

Notice how the days of the week have appeared in the left column, just like you want.

Now for the right-hand side of the Pivot Table.

Go to the builder and grab *Days* and move them to the *Row Labels* black square.

When you do this, you get the following.

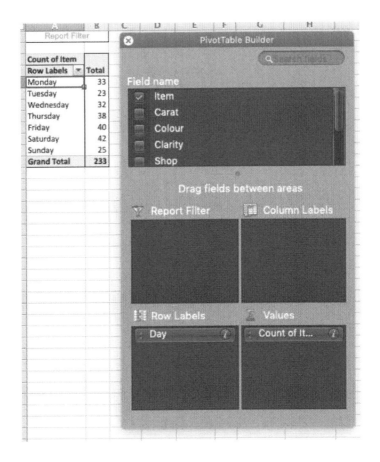

You're there.

Conclusion

This book has looked at the basic features of Excel Pivot tables.

If you wish to master Excel Pivot tables then you practice using them very thoroughly until you are very confident with these basics.

This book has thoroughly covered them.

Before you proceed further with Excel Pivot tables you must master the material in the book to the extent that you can produce pivot tables of this complexity with ease.

Once you have mastered the material in this book you are ready to become an expert with this amazing tool.

Good Luck.

Welcome to the last page reader, I'm happy to see you here I hope you had a great time reading my book and if you want to support my work, leaving an honest review will be highly appreciated. Thank you so much!

Respectfully,
William B. Skates

Made in the USA
Columbia, SC
11 March 2022

57542163R00050